The Hive Bee
A Manual of Bee-Keeping for Hawaii

by Elsie Coverley-Smith

with an introduction by Jackson Chambers

This work contains material that was originally published in 1918.

This publication is within the Public Domain.

This edition is reprinted for educational purposes
and in accordance with all applicable Federal Laws.

Introduction Copyright 2018 by Jackson Chambers

COVER CREDITS

Front Cover
*Charlie Brandts, a White House carpenter as well as
beekeeper collects the first batch of honey*
by *The White House*
[Public domain],
via Wikimedia Commons

Back Cover
Honey bee - Flickr by *gailhampshire* from Cradley, Malvern, U.K
[CC BY 2.0 - http://creativecommons.org/licenses/by/2.0],
via Wikimedia Commons

Research / Resources
The Top 6 Raw Honey Benefits
https://www.healthline.com/health/food-nutrition/top-raw-honey-benefits#1
via HealthLine.com

Wikimedia Commons
www.Commons.Wikimedia.org

Many thanks to all the incredible photographers, artists,
researchers, and archivists who share their great work.

PLEASE NOTE :
As with all reprinted books of this age that are intended to perfectly reproduce the original edition, considerable pains and effort had to be undertaken to correct fading and sometimes outright damage to existing proofs of this title. At times, this task can be quite monumental, requiring an almost total rebuilding of some pages from digital proofs of multiple copies. Despite this, imperfections still sometimes exist in the final proof and may detract slightly from the visual appearance of the text.

DISCLAIMER :
Due to the age of this book, some methods or practices may have been deemed unsafe or unacceptable in the interim years. In utilizing the information herein, you do so at your own risk. We republish antiquarian books without judgment or revisionism, solely for their historical and cultural importance, and for educational purposes.

Self Reliance Books

Get more historic titles on animal and stock breeding, gardening and old fashioned skills by visiting us at:

http://selfreliancebooks.blogspot.com/

introduction

Here at **Self-Reliance Books** we are dedicated to bringing you the best in *dusty-old-book-knowledge* to help you in your quest for self-sufficiency and food independence.

We're so pleased to bring you this old title on Apiculture. Not only is raw honey sweet and divine-tasting, it also has numerous health benefits, including antimicrobial and antiviral actions, wound-healing properties, and is packed with phytonutrients and antioxidants.

Raw, organic, locally-produced honey is also said to help alleviate the pain and suffering of seasonal allergy sufferers - a great item to have in your survival pantry when *Sneezing Season* begins!

This special edition of **The Hive Bee – A Manual of Bee-Keeping for Hawaii** was written by Elsie Coverley-Smith, and first published in 1918, making it a century old. The book is aimed specifically at novice Bee-Keepers in Hawaii.

The book has sections on *Nature and Biology of the Bee, Bee-Keeping in Hawaii, Bee-Keeping Practice,* and *The Commercial Apiary.*

A super-short, informative fast read for all Hawaiians and Hawaii residents considering taking up the art of Apiculture.

~ Roger Chambers
State of Jefferson, March 2018

English: Bee macro
Date 18 July 2015, 10:35:52
Source https://www.flickr.com/photos/55293400@N07/20118812536/
Author Ömer Ünlü
Wikipedia.com - Creative Commons Attribution 2.0 Generic

Description: Honey bee
Date: 4 November 2011, 12:55
Source: Honey bee
Author AJC1 from UK
Wikipedia.com - Creative Commons Attribution 2.0 Generic

THE HIVE BEE
A Manual of Beekeeping for Hawaii

By E. C. SMITH,

Manager Garden Island Honey Company.

INTRODUCTION.

The hive bee is a familiar example of an insect under domestication. The history of its domestication is lost in antiquity; for the art of bee culture has been known since the earliest times. In the wild state bees range the woods, nesting in hollow trees, in crevices in rocks or in any place that will afford them protection, often in the most inaccessible places. Searching for these nests in order to plunder their stores of honey constitutes a favorite pastime in the countryside. This method of procuring the bee's products, however, is wasteful and uncertain. The bee has small place in human economy until its cultivation as a domestic animal insures a lively and continued interest.

Under domestication the bee has received considerable attention. Its cultivation among the historical peoples dwelling on the shores of the Mediterranean has led to the recognition of well-marked races, from different characteristics of size, color, behavior, etc., and some improvement in quality of the different races has been made through the application of well known principles of breeding. As a domestic animal also it has been carried with the stream of human migration to practically all parts of the earth.

The German or black race of bees was the one commonly propagated by the early beekeepers and received a wide distribution; but more recently the superiority of the Italian or golden race has become recognized and it is now largely replacing the German bee. The early spread of the black bee is deplored by many beekeepers on account of the trouble it gives when they attempt to maintain pure breeds of the golden races. The principal characteristics of the Italian bee are its high color, great energy and peculiar gentleness. The queen is frequently a bright golden on the abdomen, workers are not so brightly colored and generally have five distinct yellow bands, while drones vary from very light bands to bright golden; the head and trunk in all are darker hued.

Great progress has also been made, in modern times, in the manipulation of bees. In place of the early wasteful means of securing the bee's stores by destroying the hive, which was but one step in advance of the practise of robbing wild bees' nests, has come a humane and economical practise, which has been the means of elevating bee-keeping from a simple rural pursuit and farm adjunct, giving small returns for the money and time invested, to a highly profitable industry. This revolution in the status of beekeeping dates from the adoption of the movable-frame hive, but the most advanced strides have been taken in recent years, in new countries, where beekeeping on a large scale is made possible by accessibility to extensive tracts of land and an absence of the keen competition existing in closely settled regions. At the present time industrial beekeeping overshadows every other interest in bees and this is especially true in these islands, where a large proportion of the bee colonies are owned and operated by corporations.

NATURE AND BIOLOGY OF THE BEE.

The nature and biology of the bee are of considerable interest, and a thorough knowledge of the biology is indispensable to the beekeeper who would succeed with a few or many colonies.

In the systematic classification of insects, the bee is included in that large group designated hymenoptera, on account of the common possession of two pairs of membranous wings, mandibulate mouthparts, and a modified ovipositor fitted for piercing, boring or stinging. The members of this group generally display in development what is known as complete metamorphosis, that is, growth, from eggs, is marked by immature stages known as larva and pupa, which are worm or grub-like in form and soft bodied. In addition to the bees, the group includes wasps, hornets, ants, sawflies, ichneumons and a host of related forms. The hive bee also has social

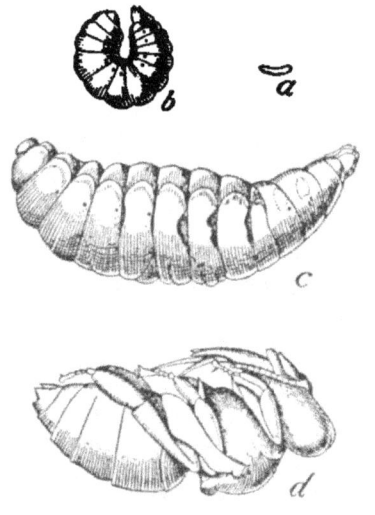

Fig. 1.
Development of bee from egg to adult (copied from Phillips).

PLATE I.

Fig. 1. Hive bees. a, worker; b, queen; c, drone.
(Twice natural size; copied from Phillips.)

Fig. 2. Colony of bees opened.
(Courtesy of Hawaii Experiment Station.)

PLATE II.

Fig. 1. Frame showing the building of comb.

Fig. 2. Frame showing brood comb.
(Courtesy of Hawaii Experiment Station.)

PLATE III.

Fig. 1. Frame showing brood and honey cells, open and capped, and the natural transition from worker to drone comb.

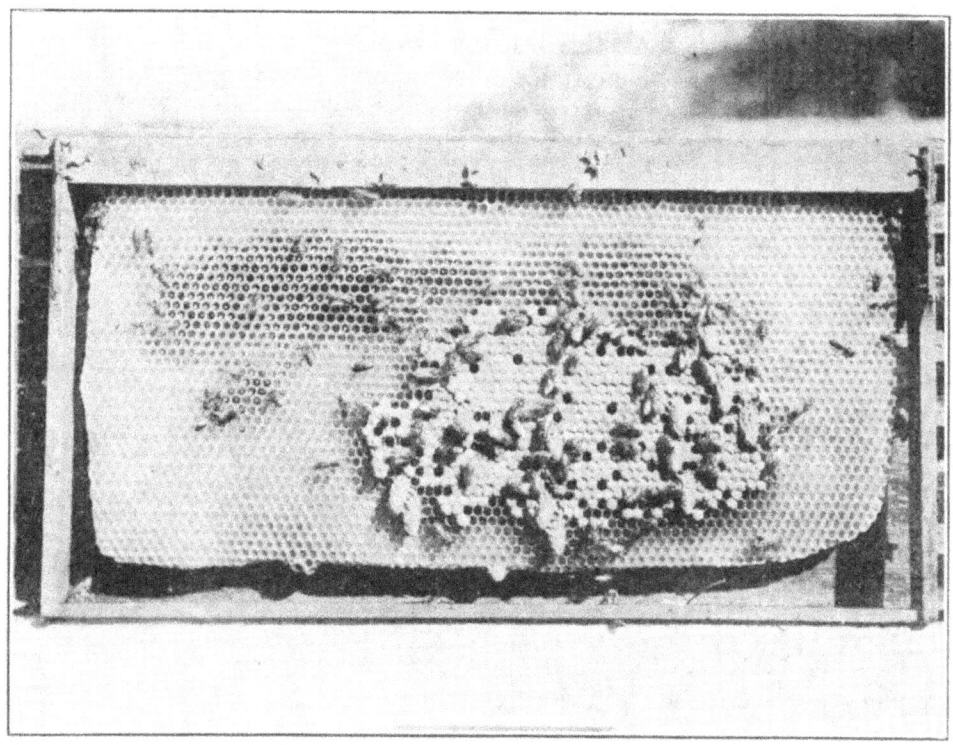

Fig. 2. Frame showing queen cells produced naturally.
(Courtesy of Hawaii Experiment Station.)

proclivities, that is, it lives in societies or colonies in which there is differentiation in form and structure among the individuals (we distinguish queens, drones and workers), and among these so-called castes there is a division of labor. A somewhat similar development has taken place in certain wasps, like the yellow jackets, bees, like the bumblebees, and in ants, whose crowded nests are familiar to almost everyone.

The biology of the bee is somewhat as follows: A colony consists of forty to eighty thousand individuals. Most of these are workers; there is usually only one queen, and at certain times there is a smaller or larger number of drones. The combs comprising the nest are pendant double sheets of hexagonal prismatic cells arranged with the long axis nearly horizontal, opening outwardly and with the angular bases interplaced and forming a common septum. The size of the cells varies from one-fifth to one-fourth of an inch across the face and from seven-sixteenths to five-eighths of an inch in depth, although the extreme sizes merge into one another with almost imperceptible gradation. The smaller ones are known as worker cells because worker bees, when raised, are always developed in them, and the larger ones are called drone cells for similar reasons, but when not used for brood, both kinds are used for housing stores. There is also another kind of cell usually found at the bottom, sometimes on the sides of the comb, larger than the ordinary cell, built perpendicular and from without much the shape of a peanut. These are actually amplified worker cells, which are built especially for the development of queens.

The queen and the drones in the bee colony represent the sexually mature forms developed by most insect species for the propagation of their kind, while the workers are a development of the social habit, and in the division of function which has followed the adoption of a social existence the labor of providing for the upkeep of the purely physical necessities of the colony falls upon her. She feeds the brood, constructs the comb, elaborates the wax out of which it is built, cleans the hive, attends the queen, repels enemies, and gathers the

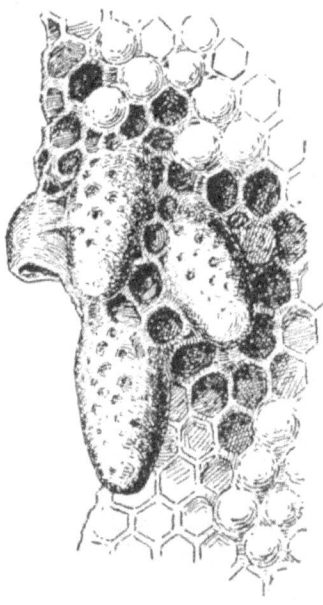

Fig. 2. Queen cells.
(Copied from Phillips.)

nectar and pollen, etc. The worker is the most highly developed of the three forms and beyond question controls all the activities of the colony.

The queen, or sexually developed female, usually within seven to ten days after maturing (exceptionally from three to fifteen days), leaves the hive and flies into the open air to become mated with a drone. This flight usually occurs from two to four o'clock in the afternoon of a bright day. The drones also are generally flying at this time, although most of the day they remain inside the hive. She returns in from ten to thirty minutes with a part of the genitalia of the drone inside and partly protruding from the hind end of her body. The drone perishes in the act of copulation. The queen mates but once, although she may leave the hive a number of times and on successive days before becoming fertilized. When she is fertilized a small sac within the body, leading from the oviduct (the spermatheca), is stored with millions of spermatozoa, the male element which fertilizes her eggs, and she is capable of laying eggs so fertilized for several years. If on a flight she is actually fertilized she usually begins to lay eggs within from one to three days. The eggs are usually placed on end at the bottom of the cells, to which they are glued fast with a sticky fluid secreted at the time of deposition. These eggs hatch in three days, and they appear at the bottom of the cells like minute streaks floating in a milky transparent jelly. The jelly is elaborated by the young worker bees of the hive. The immature bee larva feeds on the jelly, which is constantly supplied to it, apparently swallowing it through the mouth, or as has been suggested by some writers, absorbing it by osmosis through the skin. It grows very rapidly in size and occasionally sheds its skin. On the third day of its larval existence it is weaned from the jelly and put on a subsistence of honey and pollen. It is soon large enough to fill the bottom of the cell when curled up; later, to accommodate itself in the cell it must stretch lengthwise, head directed outward. On the fifth day it has gained its full growth and the cell is capped with a mixture of wax and prepared pollen. The larva within then spins a silken cocoon and its transformation into a pupa begins. This is usually accomplished on the ninth day; later, with the formation of the legs and antennae beneath the enveloping skin, the pupa begins to assume the appearance of the adult insect. When the structure of the adult bee is fully formed, the envelope is burst, the bee begins to harden and obtain its color, and on the twentieth day begins to eat its way through the capping. This is the development of the worker bee, and under ordinary circumstances the workers constitute ninety-nine per cent of the individuals of the hive.

The development of the other two castes differs somewhat in detail and they are produced only under the unusual conditions narrated below. If a queen lays before or immediately after mating, without mating at all, or after the supply of spermatozoa in the spermatheca has become exhausted, or if a worker bee should lay, it is evident that these eggs will not be fertilized. The bees which develop from infertile eggs are always drones. This phenomena, reproduction without fertilization, or parthenogenesis, is not uncommon among insects, but the results are not always the same as with the hive bee. Drone larvae can usually be recognized by their larger size and bulkier appearance. Their development also takes longer. Usually the eggs hatch in three days, the cells are capped on the seventh day, and the adult emerges on the twenty-fourth day from the egg. The cappings of the cells, too, are usually convex. Under certain conditions, in strong hives with plenty of stores, drones are produced in large numbers from the eggs of a regularly mated queen that is still capable of producing fertile eggs. As they are always produced in large-sized cells, the idea has been advanced that the contracted mouth of the worker cells regulates the mechanism of fertilization.

The conditions under which queens are produced are: the disproportionate growth of the colony necessitating division or swarming, the approaching sterility of the old queen, or its accidental loss. As a queen once mated is capable of reproducing for several years, and there is only one queen required in a hive and usually only one tolerated, it is evident that the exigency under which queens are produced is not frequent under natural conditions. When the need for the development of a queen arises, the workers break down the wall of a cell in which a fertile young larva (if there are none of these present the colony perishes) is present and begin the construction of the large-sized, peanut-shaped cell previously described. In these cells, for usually a number of them are begun at the same time, the developing larva is not weaned on the third day, but is supplied with the jelly given the young larva of ordinary bees in large quantities continually throughout its larval existence. Perhaps on account of the abundance and richness of the food, the development is quicker than with the other castes. The cell is capped on the 5th day, the larva pupates in seven days more, and the mature queen is ready to emerge on the 14th or 15th day (from the egg). The first queen to emerge usually destroys the remaining queen cells. In this she is assisted by the workers. Should several emerge simultaneously a battle ensues, from which one generally issues triumphant. Occasionally the colony is divided into two or more parts, a part of the bees swarming out with each of the surviving queens until only one queen remains in the hive. As a vagary of bee behavior occasionally two queens are tolerated and live together peaceably in the same hive. The function of the queen and drone is limited to propagation, and for this work a single mated queen is all that is required for the most vigorous colony.

The drones upon emerging from their cells wander aimlessly about the hive, being fed by the workers. They generally fly out for a short period each day, usually in the early part of a sunny afternoon. (The young queen usually takes her mating flight at the same time of day.) A hive containing many drones will be readily noticed on account of the peculiar loud buzzing noise that they produce in flight, which is very distinct.

The natural economy of the colony makes ample provision for the successful mating of the queen, but one occasionally fails to become mated and usually such queens die in the hive —whether starved or killed outright is not definitely known. Sometimes they live and after three or four weeks begin to develop like a mated queen and lay eggs. These eggs being infertile produce only drones. The colony in either case dies out. From the above it will be readily seen that a mated queen is absolutely essential to the continuance of the colony.

As already pointed out, in the natural economy of the hive, drones are usually produced in numbers when needed. As they are entirely non-producers and at the same time large consumers of the stores of the colony, as soon as their usefulness ceases they are ruthlessly slaughtered.

The workers on emerging from their cells begin to feed, and as soon as they have filled themselves with honey and become oriented to the hive, start about their work of taking care of the developing brood, which they attend with assiduous care, literally bathing them, for the first three days, as already stated, in a jelly which is supposed to be elaborated from certain glands of the head; later in a regurgitated mixture of honey and pollen slightly altered by digestive juices. The feeding of the brood and the production of wax for cell material is looked after largely by the young bees. The wax itself is secreted from glands over the inner faces of the ventral plates of the abdomen, appearing in the form of thin projecting plates after collecting on peculiar membranes in connection with these plates. In producing it the bees hang within the hive in festoons or curtains. As the bees become older, they go out and gather nectar, pollen, or propolis, they act as guards, remove the debris, and attend to any other work of the hive. With the exception of the store-gathering, however, none of these duties takes the exclusive and continuous occupation of one bee or set of bees. Any bee may take up any part of the work at any time and leave it at any time; but there is always something to be done, and all the time the bees are working strenuously, doing a little here and a little there, and by their aggregate effort the work is somehow accomplished. As previously stated, the worker is not sexless but merely undeveloped. In a colony which has lost its queen and is left without eggs or young larvae from which to develop one, it often happens that a band of workers, becoming in some unaccountable way sexually developed or capable of laying eggs, begin to lay in the cells. These eggs, of course, are infertile and produce only drones. The presence of laying workers is easily detected, as their eggs are layed without regularity, being deposited on any part of the cell and often accumulating in a pile. They sometimes hatch, however, and for a while several larvae will develop together in a cell. One after another disappears until but one remains, which matures in the usual way.

Such in a general way is the life history of the hive bee and its economy as a social organism. A consideration of some of the external factors having a bearing on its development will conclude the treatment of the biology.

First of all the bee requires a constant supply of food. The natural food of the bee, as must have been gathered from what has preceded, is nectar and pollen, which are gathered from the blossoms of many different plants; nectar also from certain plants with extra-floral nectaries, and the saccharin exudation of certain insects (leaf-hoppers, plant lice, coccids, white flies, etc.). They also require a certain amount of water, which of course is available everywhere. As the nectar and pollen are required not only by the mature bees but also by the developing brood and are obtainable more abundantly at certain times of the year than at others, the bees gather as much as they possibly can and store the excess in the hive to be used as needed and as a provision against times of scarcity. In storing the pollen in the cells it is packed in tight and the cells are left uncovered. The honey, however, after it has ripened by the evaporation of its water, is capped over with a thin covering of wax. (In this climate it frequently occurs that honey is capped before it is properly evaporated, with the result that fermentation sets in, the caps are burst, and the honey runs out of the cells and trickles down through the hive).

The existence of the colony is not only dependent on a laying queen and a continuous supply of food but there is also an optimum temperature for the brood, which must be maintained constantly lest it should become chilled and die. The bees do this by clustering over the brood, thus keeping it warm with their bodies. On the other hand, an overpopulated hive may become too warm for its occupants, in which case a portion of them will move outside while those within attempt to cool the air by fanning their wings. Should it become too cold the bees conserve their heat by clustering.

The bees must also cope with numerous enemies of the predatory and parasitic kind, which are several bacillary diseases, usually known as foul brood, the wax moth, ants, cockroaches, and other foes. The development of the sting and the liberal use in the hive of propolis (a mixture of wax and the resinous exudation of plants) are more or less successful means of repelling some of the unwelcome intruders.

The colonies must also multiply and disperse. This under natural conditions is accomplished by casting off swarms, or swarming. What the actual conditions are which impel the bees to swarm is not clearly known. It is pointed out by Phillips that swarming is always accompanied by an unbalanced condition of the brood chamber in regard to the age of the bees there. Whatever the reason, we know that the practical effect of swarming is a division of the colony which puts both parts in somewhat better circumstances and brings about a multiplication of colonies and a general dispersion of the species. The swarm consists of a queen and a large number of worker bees. After a longer or shorter period of restlessness on the part of the bees, they gradually acquire a common impulse to move and leave the hive in a body, flying about in the air until they find a convenient place to alight. Here they rest for a while in a cluster about the queen; later the cluster is broken and the bees in a body fly away to a nesting place, usually in a crevice in the rocks or in a hollow tree. When the swarm alights near at hand it can usually be hived, but many swarms are lost, some without being seen.

PLATE IV.

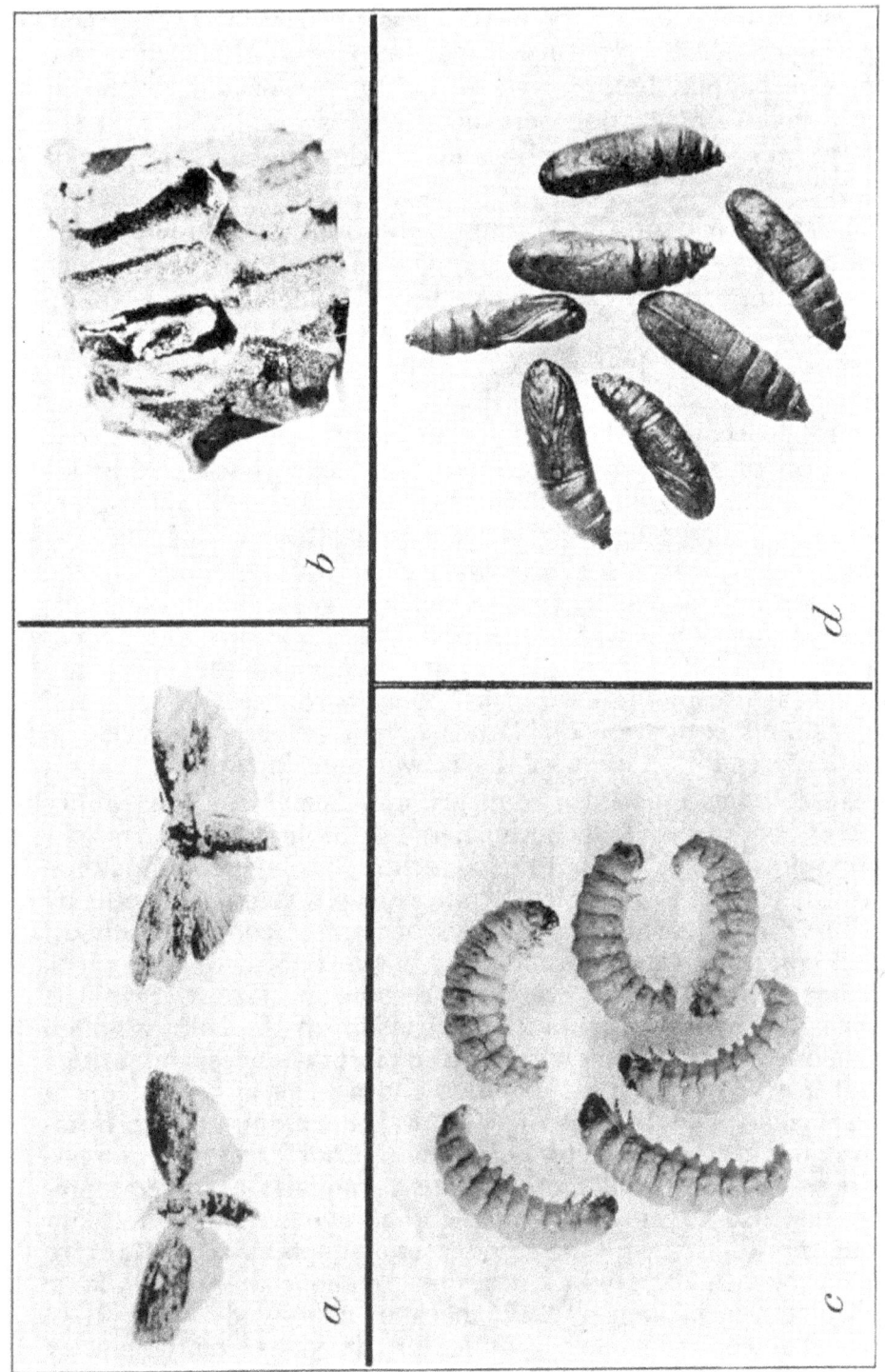

Bee moth. a, adult moth; b, eggs on comb; c, larvae; d, pupae.
(Courtesy of Texas Agricultural Experiment Station. Photo by F. B. Paddock.)

PLATE V.

Fig. 1. Comb and foundation destroyed by wax worm.
(Photo by F. B. Paddock, Texas Experiment Station.)

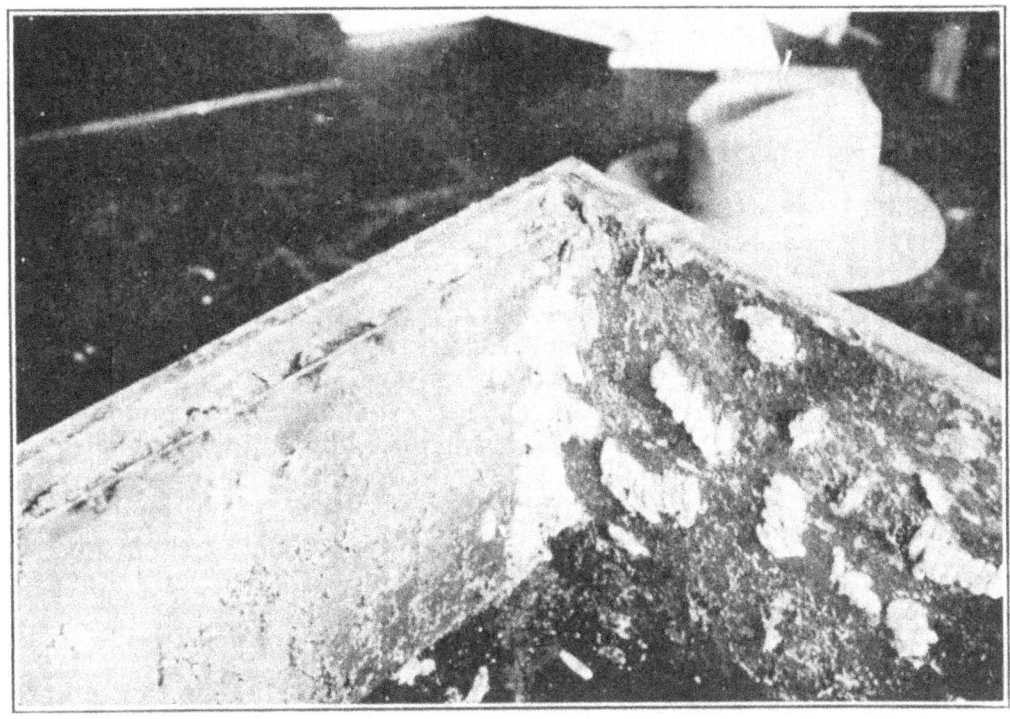

Fig. 2. Characteristic appearance of cocoons of the bee moth.
(Photo by F. B. Paddock, Texas Experiment Station.)

BEE-KEEPING IN HAWAII.

As already indicated, bees are usually cultivated for their honey, a delicious and wholesome article of food, for which there is a constant demand in the grocery, confectionary and baking trades; the wax combs can also be rendered into beeswax; and the bees themselves have a value as an article of trade.

Almost any one of intelligence can keep bees; the manipulation of the boxes requires no great exertion of strength, and the ways of the bees are readily learned. From one or two to half a dozen hives, however, are about all that can be managed properly in one's leisure time. To go into beekeeping more extensively would require the expenditure of considerable time and money, perhaps more than could be spared by the individual in ordinary circumstances. Beekeeping as a business venture, whether for the employment of capital or as a means of gaining a livelihood, is quite another matter from the keeping of a few colonies. On account of the large amount of capital required to start an apiary on a paying basis, some knowledge of the business and considerable experience with bees are almost indispensable, and a beginner is usually advised to go slowly and cautiously. It should be realized at the outset that there are many factors in industrial beekeeping beyond the successful management of bees, although this, of course, is of primary importance.

Hawaii offers a great many advantages to the beekeeper; an almost continuous season, absence of brood diseases and other serious enemies of the bees, some exceptionally good pasturage, and fairly efficient and cheap labor. The real drawback at the present time is the narrow field for extension; the small size of the territory and the previous occupation of nearly all the good pasturage.

The history of apiculture in the Islands is remarkably interesting. The first hive bees were introduced from California in 1857, after several unsuccessful attempts to bring them around the Horn from New England. These were undoubtedly black bees and from all accounts they flourished and many colonies went wild. For three or four decades following their introduction bees were kept in a very desultory fashion and without much interest· In the meantime, the rapid spread of the "kiawe" (algaroba) and the cultivation of cane had made a wonderful change in the bee forage, and it was not long before the commercial possibilities were realized. The first large apiaries were started about fifteen or twenty years ago, and since then the business has steadily grown and gone into nearly every field. Some territory is still open; other sections are only half stocked, but the advantage of closeness to market and business centers has caused the leeward side of the Island of Oahu to be severely overstocked, and several apiaries suffer from competition.

The present yearly output is about one thousand tons of honey and twenty-five tons of wax, from approximately twenty thousand colonies. About four-fifths of the total number of hives are owned and controlled by four beekeeping corporations; the remaining colonies are scattered among small apiaries. Most of the corporation-owned apiaries are managed by an experienced white beekeeper, but practically all the manual labor is performed by Japanese, who, as a rule, show great proficiency in handling bees, and who almost alone are capable of the arduous work. The smaller apiaries are usually conducted without help and as a rule are rather poorly kept, but where the Japanese can be profitably used in the business they are usually employed for the manipulation. Some Japanese have become very skilful at queen-rearing. There is only one company (The Garden Island) engaged in commercial queen-rearing.

The source of Hawaiian honeys and the character of the forage will be rather surprising to anyone unacquainted with the Islands. More than half the honey produced here is of the honeydew type, running from light amber to dark brown, according to the degree of admixture of honeydew with the floral nectar, and especially characterized by the molasses-like taste and the darker color. The principal forage plant is unquestionably the "kiawe" or algaroba, as indicated by the heavy crop produced at the time of "kiawe" bloom. Many native and introduced trees, shrubs, and herbaceous plants, notably lantana, guava, oi, ilima, koa, ohia, eucalyptus, etc., produce more or less of the honey, but the principal supply comes from the source above mentioned.

While the tropical conditions present so many features favorable to beekeeping there is one feature which is decidedly adverse, namely, insect pests. Hawaii fortunately has no contagious brood diseases, and there exists a hearty co-operation among the beekeepers to prevent the introduction of any. The insects, however, are often extremely difficult to deal with, especially where beekeeping is practised on a large scale, although a large part of the losses are more directly due to carelessness and inattention. The wax moth and ants are the worst foes of the beekeeper, but cockroaches also get away

PLATE VI.

Fig. 1. Removing frames. Fig. 2. Beekeeper equipped for work. (Courtesy of Hawaii Experiment Station.)

PLATE VII.

Fig. 1. Warehouse in connection with apiary. Fig. 2. Hawaiian apiary in kiawe grove.
(Courtesy of Hawaii Experiment Station.)

with large quantities of wax and honey in depleted colonies and abandoned hives. Ants are particularly hard on weak colonies and at times make inroads on populous hives. Where ants are bad and serious losses occur it is best to protect the hive by raising it on stands with metal legs which must be kept coated with axle grease. The wax moth is very troublesome in neglected apiaries, going through the combs in abandoned boxes very quickly. It also occasions heavy losses where empty frames are stored for future use. The spoliation is not possible where the combs are cleaned up thoroughly throughout the apiary and renewed with comb foundation when needed again.

BEE-KEEPING PRACTISE.

From the data already presented anyone interested in the cultivation of bees can understand the prospects in following this pursuit here. Cautiousness is advised to the uninitiated beginner, for much has to be learned by experience. There are many good books on beekeeping which would be helpful to the beginner, but a great deal that has been written on beekeeping practise applies more particularly to beekeeping under conditions quite different from ours. For this reason a study of the methods used here has been made with the idea of presenting them to anyone taking up beekeeping for the first time in Hawaii.

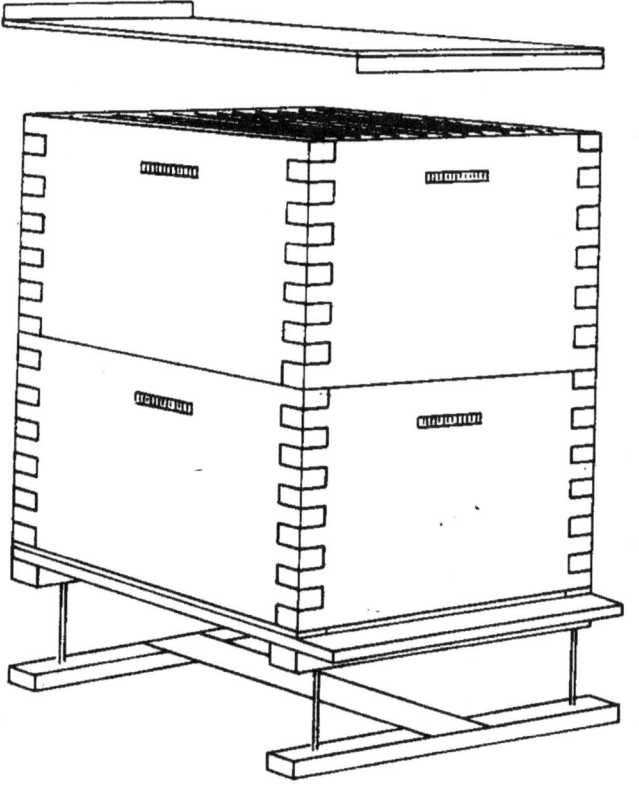

Fig. 3. Root-Langstroth hive adapted to Hawaiian conditions (original).

Fig. 4. Smoker (copied from Phillips).

To begin, one must have hives and bees and a good location for them. It is usually considered that the best time to begin is in the Spring and previous to the honey flow, but under Hawaiian conditions one can begin at almost any time without running much risk of loss or failure. The usual way to acquire bees is to purchase them from a neighboring beekeeper, and it is most convenient to seller and purchaser to deal with nuclei, or five-framed colonies with the frames well covered with bees and at least two frames of brood. While beehives may be made at home, they can be bought from dealers for much less than they would cost home-made, and the standard ten-frame hive with one or two supers, is recommended for general use. A coat or two of paint applied to the hive bodies before use preserves the wood and makes them much more durable than they would otherwise be. Select mated queens are very essential to the prosperity of the colonies, and if not acquired with the original hive they should be purchased from some reliable queen-breeder and introduced.

Since the location of the bees depends so much on the circumstances of the beekeeper, little can be suggested in this regard. One must consider primarily the quantity and character of the forage and the proximity of neighboring apiaries. The early sunlight is desirable, also a shelter from wind. The best locations in Hawaii are undoubtedly in the kiawe belt, but other locations are often quite profitable. With the best possible forage it is not profitable to put more than two hundred hives in an apiary, and apiaries of this size should be located not less than two miles from each other.

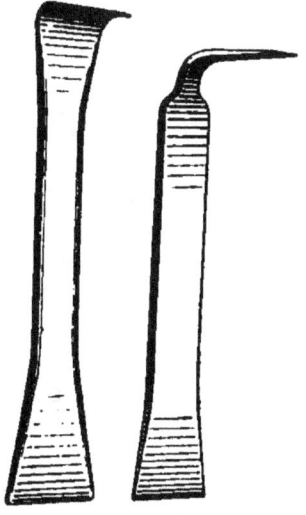

Fig. 5. Hive tools (copied from Phillips).

In addition to the bees and hives one will need at the beginning a few supplies, such as a smoker, a bee veil, a hive tool, some sheets of comb foundation, queen cages, etc.

The hives should be arranged far enough apart to allow of easy movement of the beekeeper between the hives without disturbing the bees in flight. Shade is desirable if it can be had; the hives are well placed in a grove of kiawe trees, if the latter are of more or less open growth. Accessibility should also be considered, that is, the apiary, if large, should not be located too far away from the road, and yet not so near that any animal passing is likely to be stung by the bees.

Some other ways of acquiring bees should be mentioned. The chance discovery of a strong swarm of bees often leads one into beekeeping. The successful hiving of a swarm depends somewhat upon its location. Usually the swarm alights on the branch of a tree, and in this case the limb can be cut down and the bees shaken off in front of a hive previously prepared for their habitation. If the swarm alights high, or it is not desired to sacrifice the branch on which it rests, the bees can be shaken into a bag on the end of a pole and the bag emptied before the hive, or the bees may be emptied into the hive. In preparing a hive for the bees it is always advisable to place in it one frame containing honey and some very young brood. Under these circumstances the bees never desert their home and are in a position to provide for their need in case anything has happened to their queen.

It is a very common occurrence in this country for bees to build their nest in the walls or ceiling of some one's home. This is very annoying to the inhabitants and yet the bees can be removed without damage to the home and at very little expense to a beekeeper, in the following way. Take a bee hive and place in it one or two combs containing both young brood and honey, also one or two sheets of comb foundation, a queen bee and a handful or two of working bees. Place the hive thus prepared so that the entrance to the hive is close to where the bees enter the house. If the bees have several exits from the house all but one must be blocked and that one fitted with a bee-escape. If the work is done well the bees can get out but cannot return and each bee as it goes out to work not being able to get into the house will enter the hive. It will take from three to four weeks to get the last of the young bees out but when all are at last out the bee-escape can be removed and the much strengthened colony returned to the apiary. At the same time, the performance of a neighborly act brings a mutual benefit.

It is also possible at times to purchase colonies in box hives very cheaply. As these can be readily transferred to standard hives it may be considered good business to buy them. To transfer bees from a box to a standard hive, take the box a few feet from where it stood and put the standard hive in its place; see that the standard hive has sufficient frames with comb foundation to accommodate the bees, also a comb containing a little honey and very young brood is advised. This being done, remove the top of the box hive and place an empty box without bottom over it. Tap soundly on the sides of the lower box with a stick and the frightened bees will start to move into the upper box. After a time they will all be clustered in the upper box and can then be shaken into the standard hive. In the meantime, all bees on the wing will return to the old location and enter the new hive, and in a short time will all be at work, just as though nothing had happened. It is well to save as much of the young brood as possible. This can be cut out of the old box and placed above the queen excluder in the new hive. In this way it will be taken care of by the worker bees and in the course of two or three weeks, when the last bees have emerged from the cells, the old combs can be removed and rendered down for what honey and beeswax they contain.

In purchasing bees it is often necessary to move them from one place to another. Their transportation is not particularly difficult or hazardous, if careful attention is given to the following details. If possible a cool day should be selected and ventilation provided according to the strength of the colony and the length of time it is likely to be confined. The best method of securing the ventilation is to remove the hive cover and more or less of the frames of honey that are in the hives, fastening the balance of the frames so that they cannot get out of place. To keep the bees in, attach wire gauze over the top and entrance to the hive and, thus prepared, move with care to the new location.

If there is plenty of bloom at the time and the directions given above are carefully followed, the bees will go to work vigorously, and the frames will rapidly fill with honey and brood. As these frames are filled, more frames with comb foundation should be given, and as soon as necessary an extra hive body ought to be added. A queen excluder is always placed between the bottom box and the hive body above to prevent the rearing of bees in the hive body. The surplus honey is taken from the frames in the hive body, and the queen is excluded from this so as not to mix the brood and honey in the same frame.

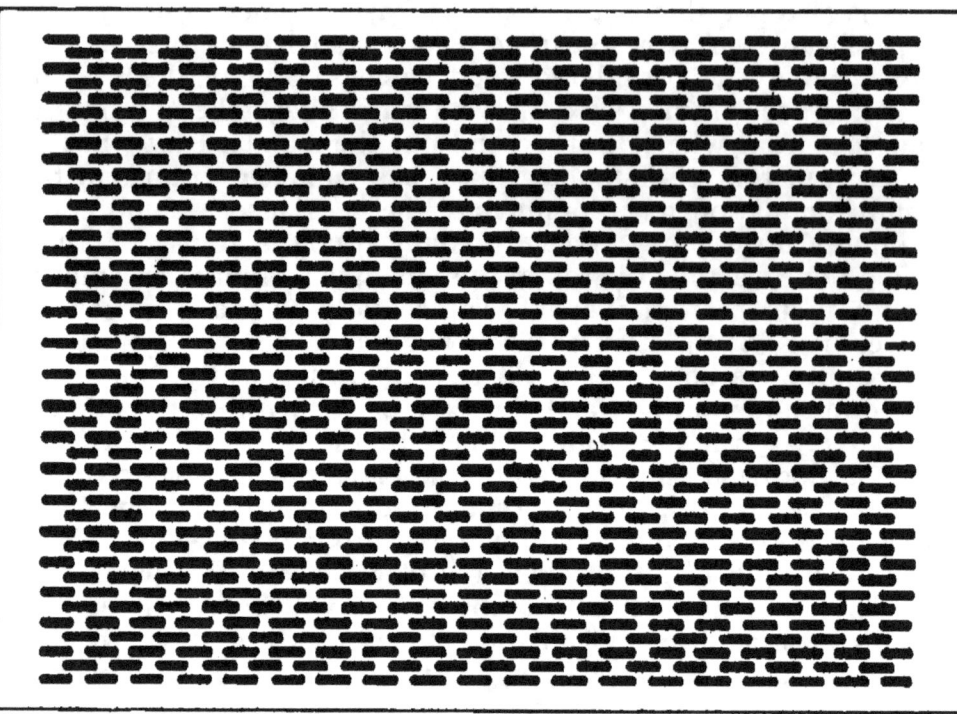

Fig. 6. Zinc queen-excluder (original).

With all the bees working and the hives filling with honey, there comes the question of disposition of the surplus. Occasional examinations of the apiary should be made so as to determine the quantity and quality of honey coming in, and thus decide what frames or hive bodies to add or when to extract so as to obtain best results. If a hive appears overcrowded with bees, honey should be taken out or extra hive bodies added.

When examining a hive, care should be taken to see that the cover is properly adjusted before leaving it, as honey exposed for any length of time in a hive is likely to start the bees robbing. It is easier to prevent robbing by exercising a little care, than it is to stop it after it is once started. Replacing the cover and greatly narrowing the entrance will stop the marauding bees if they have not gone too far. Exposed stores anywhere will often start robbing, so storerooms should be made tight, and combs with honey should not be left lying around.

As the great majority of beekeepers keep but a few colonies and as most of these colonies are kept for the sake of what honey they will produce for home consumption, this kind of bee-keeping should be made as attractive and pleasant as possible. Where the industry is so limited, it is recommended that the beekeeper produce what is known as section honey, that is, comb honey put up in small neat frames by the bees. This has many advantages for small beekeepers. In the first place, a very attractive as well as a delicious table article is secured. Furthermore, the equipment necessary for the production of comb honey is not expensive, while the work is clean, simple and light. There are two types of sections in use in these islands, namely, the plain section and the bee-way section. The plain sections are as wide as the ordinary comb of honey is thick, and when in place in the hive are spaced by fences which allow a beeway so that the bees can pass in and out at work. The bee-way sections are a little wider and have the passage for the bees cut in both sides at the top and bottom and while they need no other spacing it is best to keep them apart by the use of some thin material such as strips of zinc or thin pieces of wood. Either section makes a very attractive package when properly filled.

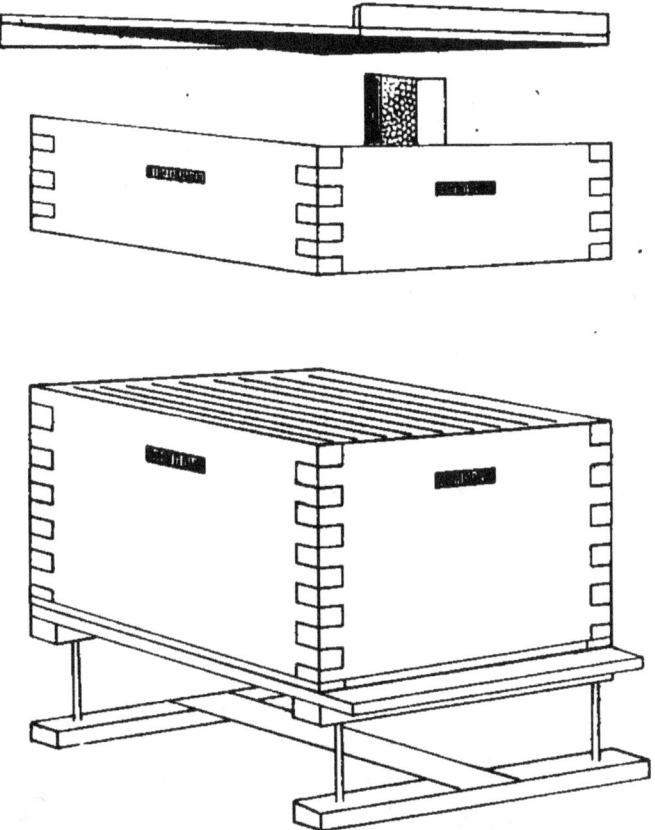

Fig. 7. Root-Langstroth hive with shallow super for comb-honey production (original).

The important commercial product, however, is extracted honey, which is produced profitably only in the larger apiaries, and for this the initial expense is great. It can be said, however, that the increase of production lessens the unit cost and so augments the profits. The limitations to the business come in the lack of profitable territory. The first requisite for extracted honey production is an extractor, which is a patented mechanical device for removing the honey from the frames by centrifugal force, after the cappings have been taken off with a capping knife. A wax press is also necessary to obtain the wax from old combs, slumgum, and miscellaneous fragments in which there is a great deal of foreign matter. It greatly facilitates the work of recovering the wax from this kind of material. The old-fashioned method of rendering wax in a boiler over a fire is still the usual way of recovering wax from the comb. If the apiary is large, it is desirable to have an extracting house as well as one for general storage purposes. In the out-apiary, the extracting is frequently done under a tent. In conducting out-apiaries, teams or an auto-truck is necessary to move the honey. Hive bodies, frames, a wheelbarrow, shovel, and other implements also go to make up the equipment. After extraction the honey is stored in bulk in barrels, cans, or tins, and may be kept thus for a considerable length of time without spoiling. A tendency of nearly all pure Hawaiian honeys is to crystallize soon after extraction. As it is difficult to overcome this tendency permanently, they cannot be considered good bottling honeys.

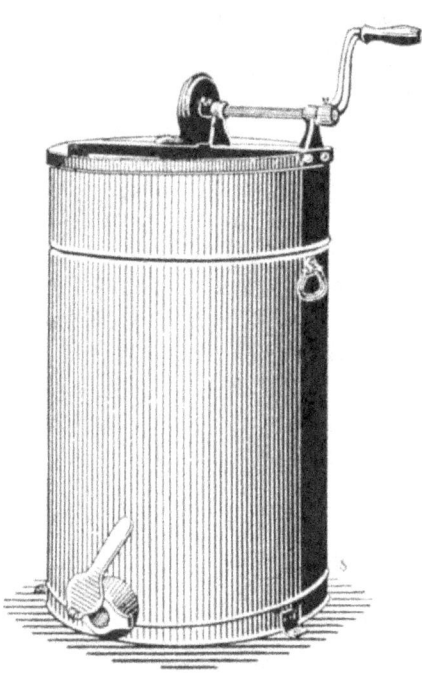

Fig. 8. Honey extractor (copied from Benton).

Since the bees will not build straight comb in the frames without the use of comb foundation, and the usefulness of the

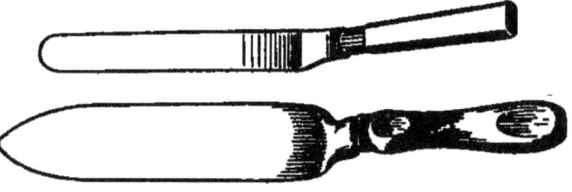

Fig. 9. Knives for uncapping honey (copied from Phillips).

movable-frame hive would be greatly impaired otherwise, a comb foundation mill is also a necessity in the equipment of a large apiary. The foundation mill is a patented device consisting of two rollers with impressed surfaces working much the same as does a clothes-wringer, in which sheets of wax are pressed out in the form of the septum between the double layers of cells in the natural comb of bees. When these are furnished to the bees the cells are drawn out rapidly from the impressed lines and comb building is greatly facilitated. Different sizes of foundation can be made, and the thickness can be regulated. It is customary to put full-sized sheets of foundation below the queen-excluder. This portion of the hive is known as the brood nest, and if care is taken in supplying the foundation for the comb, nothing but worker bees will be reared. Narrow strips of comb foundation can be and frequently are used in the supers above the queen excluder.

One of the most favorable conditions surrounding Hawaiian beekeeping is the rapid rate of increase. The beekeeper with a few hives will often be perplexed about providing for the quickly multiplying colonies, but one desiring to extend his holdings finds this phenomenal reproductivity a valuable asset in his business. Without the interference of the beekeeper the natural increase will be cast off in swarms, which, under favorable circumstances, can be hived, if it is desired to increase the number of colonies. For the professional apiarist, however, this process is obsolete and is replaced by artificial swarming or a division of the colony. Sometimes the colony is divided into two equal parts and a day or two later a queen cell is given to the half having no queen. Usually, however, the new colonies are started by removing a populous colony from its stand and putting a new hive with one or two frames of brood and one or two sheets of comb foundation in its place. The working bees, as they return from the field work, will enter the new hive and make up a second colony, and within twenty-four hours will be working as though nothing had happened, eventually rearing a queen from the young brood supplied. At this time it is well to give them a nine-day-old queen cell. The queen will emerge the next day and thus save the bees some loss of time. This style of colony is usually called a nucleus, and must build up its numbers before it can supply any surplus honey. The populous colony should be moved at least four or five yards from the old stand. Should a natural swarm occur, the opportunity to hive it should never be neglected.

This brings us to a consideration of industrial beekeeping proper, or a business running a thousand or more colonies.

PLATE VIII.

Fig. 1. Types of honey containers used in Hawaii.

Fig. 2. Refitting frames with foundation.
(Courtesy of Hawaii Experiment Station.)

PLATE IX.

Fig. 1. Comb foundation mill.

Fig. 2. Hives protected from ants. (Courtesy of College of Hawaii.)

THE COMMERCIAL APIARY.

It is characteristic of industrial enterprise to strive for a lower cost of production and in this regard industrial beekeeping makes no exception. The determining factors are economical management of labor, utilization of labor-saving devices, etc., etc. Additional apparatus is also necessary. The solar extractor, for instance, is a device for utilizing the sun's rays in melting down the wax of broken comb, burr comb, granulated honey comb, etc., and of great utility in a large establishment. This is a strong box, usually somewhat elevated from the ground, and with a removable glass cover. The box is lined with zinc, and has a false bottom at about one-half its depth. This false bottom is made of heavy wire gauze or perforated zinc supported by means of cross-bars of wood. The bits of comb and granulated honey are thrown in on this false bottom, which, acting as a strainer, allows the wax and honey to pass below as fast as it is melted, and at the same time keeps the balance close up under the glass where it gets the greatest amount of heat. The box is usually painted a dull black and when in use is tilted a little so as to get as much of the sun's direct rays as possible. Once melted the honey and beeswax separate and the clear honey is drawn off from the bottom of the box by means of a faucet provided for that purpose, and the cakes of wax can be lifted out and treated as desired.

A very efficient apparatus for rendering wax is that invented by Smith and Waterhouse and used by the Garden Island Honey Company. This device is designed to handle any kind of wax it is desired to render, but is especially useful in obtaining that which adheres to the frames after they have had the main part removed by the use of a knife. The apparatus consists of a water tank and a grease or wax trap, the latter being connected with the tank at both top and bottom by means of pipe. The pipe is fitted with a sufficient number of stopcocks for convenient manipulation, and a brick fire-box is built around the water tank to conserve heat. When in use the tank is nearly filled with water and wax. A strong fire is built underneath, which brings the water quickly to boiling, and by means of the ebullition and the formation of steam a circulation is started in the pipes. The melted wax is carried along with this movement of the water as far as the wax trap where it collects and floats on the top of the water until removed.

The use of this device in recovering the beeswax from

frames necessitates the constant refitting of the frames with comb foundation, but the value of the extra wax recovered (having in mind especially the depredations of the wax moth in old stored comb) greatly exceeds the extra expense. With a heavy flow of nectar extending over a short period there might be some advantage in having the old comb to place in the hive, but it is doubtful if a full sheet of comb foundation is not equally as good.

In all extensive bee business there is a constant demand for large numbers of queen bees, but since queen breeding is a line of work altogether apart from ordinary apicultural management and with peculiar exactions in the way of time and skill, it is a matter of opinion whether it is more economical to buy queens from a regular queen breeder or to go into queen breeding on one's own account. Under any circumstances, to secure good and economical results this work requires the constant attention of one person. The methods of queen rearing are extremely simple, when once grasped, but require a large degree of application and much attention to detail. Many methods of doing this work are in vogue: the description of one is sufficient. A few tools and appliances are necessary. A cell molder, budding needles, frames for holding queen cells, jelly spoons and queen cages, etc., form part of the equipment. The operation begins by preparing artificial queen cells on a special frame and procuring a little royal jelly from a natural queen cell. Natural queen cells are always found in a strong hive that has been without a queen for three days, so that in order to get the royal jelly it is necessary to remove the queen from a strong colony. If the queen happens to be one that is needed, take sufficient bees with her to care for her; otherwise she may be destroyed. After this the young larvae from which it is desired to rear queens should be procured. Larvae not more than one day old are best employed, and great care should be taken in their selection. Then proceed as follows: Take the budding needle and break down the walls of the queen cells found in the dequeened hive, remove the larvae or immature queens and then with the jelly spoon take just a little of the royal jelly that is found in the bottom of these cells and smear it on the inside of the artificial cell, leaving a small amount at the bottom. All cells being thus treated, once more use the budding needles to transfer the selected larvae from their natural cells to the artificial cells, and as soon as the work is completed, place the frame or frames of queen cells in the hive from which the royal jelly was taken. These frames should be placed near

PLATE X.

Fig. 1. Solar extractor. (Courtesy of College of Hawaii.)

Fig. 2. Wax machine. (Courtesy of Hawaii Experiment Station.)

PLATE XI.

Fig. 1. Queen cells artificially produced.

Fig. 2. Shipping cages for queen bees.
(Courtesy of Hawaii Experiment Station.)

the center of the brood nest to get best results, but never put two frames next to each other. If the work is well done, the majority of the larvae are usually carried through to maturity and from twenty to fifty good queens may be looked for. From the time the frames are put into the hive, care should be taken to see that a queen does not emerge from a cell that has by chance been overlooked in tearing down the natural queen cells, as such a queen would destroy all the artificial queens, breaking open their cells and eating the larvae. In six days examine the frame of queen cells and find out how many are good. If they are to take the place of old queens, destroy the old queens at once and on the ninth day distribute the cells where needed. If increase is desired, nuclei should be prepared on the sixth or seventh day after budding and the cells distributed on the ninth day. In any case the cells must be distributed, caged or destroyed on the ninth day, as the first queen will emerge on the tenth day and all other queens will then be destroyed.

In following the above directions in any extensive continuous breeding work one will immediately see the necessity of keeping accurate record of all operations, and any system of keeping records will involve the use of indubitable marks on boxes and of a book for recording all the data connected with the different operations necessary to procure queens.

It is needless to say, however, that paramount to all the care and painstaking work involved in rearing queens to maturity is the necessity of experience and judgment in the selection of the breeding stock. The qualities of the queen will depend on its heredity, and since this can be determined from but one side, greater skill is demanded in the matter of selection. In this regard, however, criteria differ, some laying stress on the capacity of the bees for work, on their gentleness, or in a word, on "utility" characters; others again on color, size, vigor, etc. In a large apiary it is often possible to combine all these attributes in one queen or strain of bees. With any queen breeding system established in good running order it is possible to re-queen all the hives in the different apiaries frequently with first-class queens, and since the productivity of the hive depends so much on the queen it is almost essential that this be done once a year or at least every second year, either before or after the heavy flow of nectar.

Often in industrial beekeeping where low grade honeys are principally handled, the wax production becomes extremely important. This has brought about a practise known as "shaking," the main idea of which is to stimulate comb building. This is undertaken only in a limited way, and as follows: The top box or boxes are removed and all the frames in the bottom box examined. Only those containing a large amount of brood are left, and among those remaining (usually seven are left) are interspersed full sheets of comb foundation. This is frequently done twice a year, so that six combs of a ten-frame hive are recovered and rendered into beeswax during the year. The bees do not suffer from this treatment and it seems to prevent swarming. In addition, all honey and beeswax are taken from the top boxes. The practise adds much to the wax production.

A regular routine of examination, extraction, requeening and general overhauling does much toward economizing labor, systematizing the work and putting the business on a paying

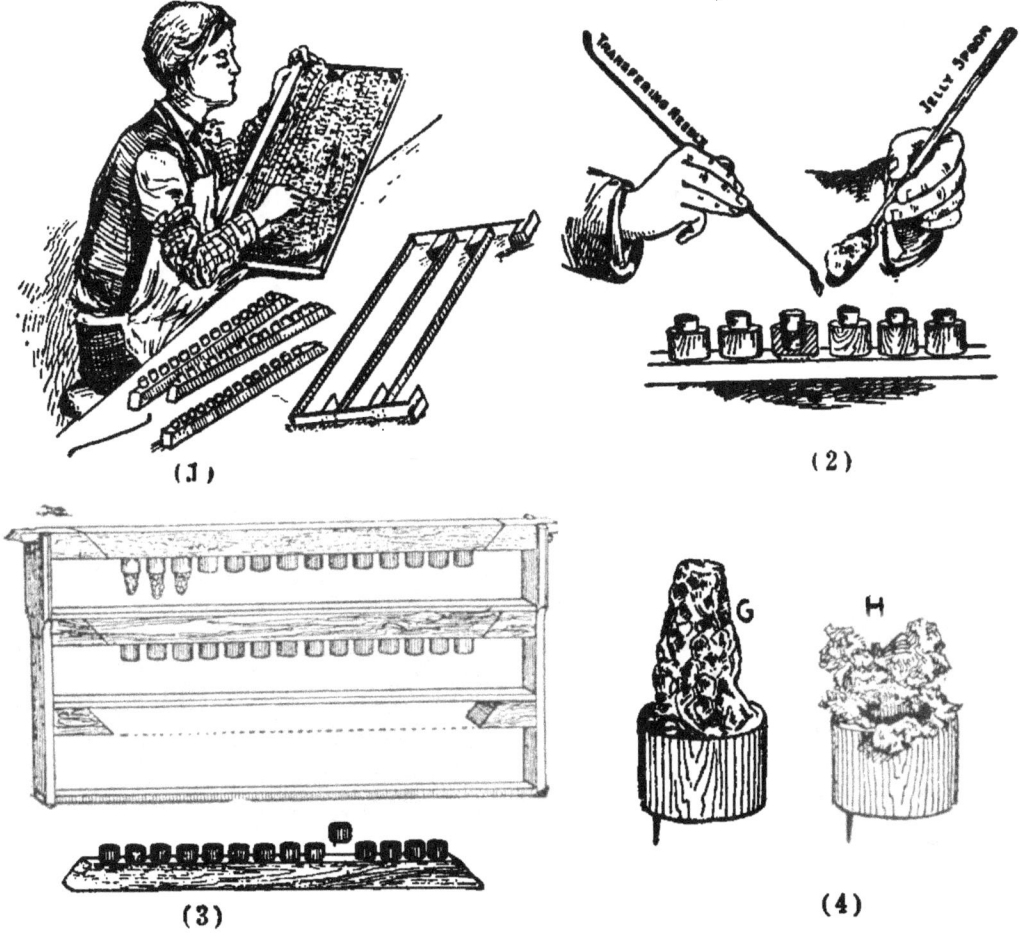

Fig. 10. Queen-rearing operations: (1) Transferring the larvae—enlarged below; (2) putting royal jelly in cell cups; (3) cell bars in frame; (4) old cells to be renewed. (Copied from Scholl.)

basis. In replacing old queens with young ones it is always best to leave the colony queenless for about two days after the old one has been killed, and the best and most economical method is to introduce nine-day-old cells. These are nearly always accepted, and once in a strong colony have every chance of proper development. However, it is necessary to examine the hive in a few days to determine if a queen is present; if not, a second one should be given. As soon as the young queen begins to lay, her wings should be clipped. This prevents her going off with a swarm, and is a mark to go by when the bees are examined the following year or at any future time. In this way, it is possible to tell how many queens have been lost in one year or two years, etc. It also makes it possible to compare the work of a one-year-old queen with that of any other queen, and to ascertain at what age a queen does her best. It is not unusual to find an old queen with clipped wings and her daughter doing duty at the same time, and on rare occasions two young sister queens may be found laying in the same hive. In the examination of the bees one occasionally comes across evidences of laying workers. As such colonies are always queenless and it is difficult to get them to accept a queen, the best procedure is to divide them, frames, honey and all, among other colonies, and if an extra colony is desired, make up a new one in the way previously explained.